SECRETS
OF THE DEEP

OTHER BOOKS BY STEPHEN SPOTTE

Fish and Invertebrate Culture
Marine Aquarium Keeping

SECRETS
OF THE DEEP

STEPHEN SPOTTE

DRAWINGS BY

GORDY ALLEN

CHARLES SCRIBNER'S SONS · NEW YORK

Library of Congress Cataloging in Publication Data
Spotte, Stephen
    Secrets of the deep.

    Bibliography p. 135
    Includes index.
    1. Aquatic biology. I. Title.
QH90.S65   574.92   75-19304
ISBN 0-684-14376-3

Grateful acknowledgment is made for permission to
quote material as follows:
on page 5, from "Heaven," from *Collected Poems of Rupert Brooke* by
Rupert Brooke. Reprinted by permission of Dodd, Mead & Company,
New York; Sidgwick & Jackson Ltd., London; and McClelland and
Stewart Limited, Toronto.
on page 63, from "The Descent of Winter," from *Collected Earlier
Poems* by William Carlos Williams. Copyright 1938 by New Directions
Publishing Corporation. Reprinted by permission of New Directions
Publishing Corporation.
on page 81, from "The Marshes of Glynn," from *Poems of Sidney Lanier*
by Sidney Lanier. Reprinted by permission of Charles Scribner's Sons.
on page 111, from "Men Have Chosen the Ice," from *Notes of an Al-
chemist*, poems by Loren Eiseley. Copyright © 1972 by Loren Eiseley.
Charles Scribner's Sons, 1972. Used by permission of the copyright
holder.

In memory of
John M. King II

Contents

CONTENTS

Preface

This is a series of vignettes on the natural history of the oceans, examining marine plant and animal life in the various zones of the earth. Although not technical, the information I present reflects accurately the most recent findings of the scientific community. In writing the book, I thought of it as a leisurely walk along a deserted shore. With no companions to amuse, I could stop when and where I pleased and examine whatever I found interesting.

Selecting the subject matter was easy. Two aspects of the undersea world have always fascinated me: the special adaptations marine creatures have evolved as aids to survival, and the intricate rhythm and balance peculiar to entire communities of living things. First, the book stresses the individual organism's approach to the problems of survival; next it describes the splendid variety of life-styles woven together. When dealing with this latter area I touch the tropics, mid-latitudes, and poles in cursory fashion, making quick forays into representative aquatic communities at each location. The effect is to cover the world's oceans in a few pages. Impossible? Not really. If a shoreline stroll is only in the mind, why confine it to a single continent?

I wish to acknowledge the help of several people who read the entire manuscript and offered many valuable

PREFACE

criticisms and suggestions. They are: James W. Atz and C. Lavett Smith, Department of Ichthyology, American Museum of Natural History, New York City; Forrest G. Wood, Marine Biosystems Department, Naval Undersea Center, San Diego, California; Laura E. Kezer, Education Department, Mystic Marinelife Aquarium, Mystic, Connecticut; and Bruce R. Powers, Department of English, Niagara University, Niagara Falls, New York.

SECRETS
OF THE DEEP

Prologue

The sea surrounds us. Our continents are mere islands of weathered granite and basalt awash on a planet of water. The sea is within us too. The blood in our veins is remarkably similar in composition to ocean water, a reminder of a time long ago when the land and air were uninhabitable, and only the sea supported life.

The first water emerged from the heat and turmoil of the young earth four billion years ago, when the planet's basic components were still being sorted. Metals and other heavy substances were sinking to the center while the lighter granite and basalt were being displaced upward.

The sorting was one of great force and upheaval. As the earth solidified, the pressure generated by gravity produced intense heat. At the core, where the tension was greatest, the contents boiled and churned until there were explosions that shattered the outer crust and spewed rivers of molten rock over the land. At such times gases were released, including hydrogen and oxygen, the components of water.

For millenniums the earth underwent violent changes. Great clouds of water vapor stretched against the sky, imprisoned there by gravity. As the vapor thickened, the rays

of the sun were blocked and the cooling process acceler-
ated.

But the earth had matured. Mountain ranges had come
and gone, the turmoil within had subsided. When the atmosphere had cooled below the boiling point of water,
the first rain fell. It rained for centuries. The water rounded
the sharp edges of the rocks and washed salts from the
land into the great basins where the oceans were born.

Part I

SURVIVAL

And in that Heaven of all their wish,
There shall be no more land, say fish.

—Rupert Brooke, "Heaven"

1. Adaptation

Three billion years ago life was born somewhere in the drifting froth of the sea. There, rocked by the earth's gentle spin, goaded and tortured by radiation from deep in interstellar space, a single pulse beat weakly where before had been only chemistry. Life did not suddenly appear. It died and was reborn countless times until a simple cell survived and reproduced. From this tentative start, a primitive plant evolved. The first plant consisted of a cell adapted to fuel itself on solar energy. I can imagine it tumbling over and over in the wash of that distant sea.

Animals came later. Unable to produce tissue from solar radiation, they solved the problem by eating plants and obtaining energy secondhand. Animals developed methods of locomotion and spread through the oceans of the world. At first they could move only by directing their cellular contents to flow in one direction, but later jointed legs appeared, and a spinal column for the attachment of paired fins. But each halting step in evolution, no matter what the purpose, was fraught with doubt; each new organism ran a gauntlet of its peers. Was a reptile that looked like a modern dolphin to survive in the selective oceanic community? It was not, and ichthyosaurs became extinct. Many other experiments were successful, giving their inventors a slight edge over competitors and thereby allowing them to survive as species and reproduce. These modifications in behavior or body form for the purpose of survival are called adaptations.

Spike mackerel schooling

2. Schooling

In spring the mackerel come to Coney Island in New York City. People gather by hundreds along the public pier to sit in the sunshine and talk and fish, if fishing is what it can be called, since any shiny little object with a hook can be used as bait. Nearly anything works, even pieces of aluminum foil, or a feather from an old hat. When the water is quiet you can watch the spike mackerel (so named because of their small size) lying underneath the pier like row upon row of five-cent cigars—the very embodiment of conformity.

Mackerel are schooling fish; that is, they mass together with others of their species in an organized fashion, moving in the same direction, keeping the same distance apart, and conducting the same activities in synchrony. How they do it, and why, still has not been answered satisfactorily, but one thing is certain: schooling is an aid to survival. Within a gathering of thousands of individuals, all looking and acting alike, the odds that any particular one will be eaten by a predator are slim. On the other hand, a sudden change in the physical appearance or behavior of one specimen destroys its anonymity and subjects it to immediate danger. In conformity there is safety.

Bluefishermen have known this for years. By positioning a boat over a school of menhaden upon which bluefish are feeding and dragging an unbaited treble hook through the teeming masses until a menhaden is snagged, they single out one, which will invariably be struck by a bluefish. A thrashing menhaden among thousands of others behaving normally is too conspicuous. One fleeting moment of individuality is fatal.

Barnacles closed at low tide

3. Attachment

Stand on a precipice of a rocky headland and look down at the crashing surf. The battering that the shoreline receives is intense and relentless. Such is the force that carves deep caves from granite and reduces boulders to sand. The continuous grinding and wash of the waves would seem to preclude all life, yet this area, known as the intertidal zone, so teems with living things that new recruits riding landward with each high tide can scarcely find a foothold.

Perhaps the most interesting thing about the intertidal zone is the great variety of mechanisms that animals and plants have evolved to keep from being washed away. Seaweeds are anchored to rocks by means of holdfasts, structures vaguely reminiscent of the roots of higher plants. The frond, or main body of the weed, is soft and supple, an adaptation allowing the plant to sway back and forth with the surge. A brittle plant would quickly break apart under the constant hammering of the waves.

The sea anemone, which is related to the sea jellies, attaches to rocks with the flat bottom part of its body called the basal disc. Barnacles use a different technique. A barnacle's body is encased in a stony shell, which in turn is attached firmly to a rock by means of cement produced by the animal itself. The curved, conelike shape of the shell diminishes the force of oncoming waves much as the pointed bow of a ship parts the sea with minimum resistance.

The blue mussel has discovered yet another way of holding onto a rock in the intertidal zone. It throws out several byssus threads, which are manufactured by the underside of its foot. The threads provide stability by working like anchor lines. Other mollusks, notably the chiton and limpet, have a strong, muscular foot. The foot acts like a giant suction cup, pliable and ready to assume the shape of any surface over which the animal is crawling in search of algae to eat.

Sea stars and sea urchins use another variation of the suction-cup principle. Instead of just one foot, they possess many hundreds of tiny tube feet. Battering surf simply tightens their grip on a slippery rock, just as a rubber suction cup sticks harder when it is pressed down.

Venomous lionfish

4. Venom

About a thousand different kinds of animals that live in the sea are venomous or poisonous. They are distributed widely, and evolution seems not to have favored any group in particular: species ranging from single-celled organisms to creatures so advanced as the fishes are all represented.

There is, of course, a difference between the terms venomous and poisonous. A venomous animal possesses specialized glands or cells, in most cases a venom duct, and a structure for transmitting the venom. A poisonous animal, on the other hand, has no venom apparatus at all. It poisons its victims upon being eaten. Thus some venomous animals are poisonous, but poisonous ones may not necessarily be venomous.

The subject here is limited to venom in fishes. Unlike venomous snakes or even cone snails, which are marine mollusks, no fish uses its venom for any purpose other than defense. These other animals utilize venom to capture or subdue prey, but in fishes the venom apparatus is never a weapon of attack.

Venom glands evolved independently among different families of fishes. This is why the form and location of the apparatus differ considerably from one group to the next. The spiny dogfish shark has two venomous spines, one in front of each dorsal fin. Stingrays have a single stinger at the base of the tail. Underneath the stinger and near the edges is a groove containing venom-producing cells. The stingray's defense is to lash out with its tail if attacked or stepped on.

Scorpionfishes bristle with venomous spines. The dorsal, anal, and pelvic fin spines contain grooves with venom glands. The stonefish of the tropical Pacific, a member of the scorpionfish family, is perhaps the most venomous fish in the world. Toadfishes and weeverfishes, commonly found along the American east coast, have stingers on their gill covers and also on their dorsal spines. The surgeonfishes that browse in dense schools along coral reefs have two stingers, one on each side of the base of the tail. These groups represent but a sampling; there are many others. How effective a device the venom apparatus really is has not yet been demonstrated. It doubtless has survival value, although in some cases this appears to be limited. In the tropics several species of stingrays prefer to live in shallow back bays. So do many sharks, and stingrays are a prime source of shark food. Sometimes a shark caught by hook and line has a dozen or more stingray spines embedded in its jaw, proof that despite their wonderful adaptation the rays were eaten anyway.

Nassau grouper changing color

5. Protective Coloration

Many creatures can hide effectively because their own coloring matches that of their surroundings, a device called protective coloration. Perhaps no group of animals anywhere has so perfected this technique as have the fishes.

Bottom-dwelling species are often masters of disguise.

SECRETS OF THE DEEP

The winter flounder, a small flatfish native to New England, can change its color instantly to match the sea floor over which it is swimming. If the bottom is light in color, so is the flounder. Winter flounders have such precise control over their skin pigments that laboratory specimens have been known to assume near-checkerboard patterns when placed in aquariums with checkered bottoms.

Fishes that live in the open sea where there is no background at all have taken protective coloration in a different direction. Those fishes with a silvery, reflective appearance secrete nitrogenous compounds called guanine and hypoxanthine, which form thin crystals in the scales and skin. Batteries of these crystals, each one so small it can be seen only through an electron microscope, reflect available light like millions of tiny mirrors.

The crystals are arranged in stacks composed of alternating layers of crystals and cytoplasm. Each stack, called a platelet, reflects all light that penetrates it. Even though some of this light may be green, red, or some other color, the total contains all colors of the spectrum and looks white, or silvery. A silvery fish viewed in a horizontal plane is nearly invisible because the light reflected by the platelets in its sides is virtually of the same intensity as the light in the background.

A fish's back presents a different problem. Here the objective is to reflect as little light as possible because the sea is dark when viewed from above. Most species have solved the problem by evolving heavily pigmented backs. Platelets on the back lie parallel to one another with the edges pointed up, allowing light to pass between them to dark skin pigments below.

The part of a fish most difficult to hide is its belly. Because the surface of the sea is normally brighter than the fish when the animal is viewed from below, the fish's outline is silhouetted. To counteract this effect, many species have evolved light-colored bellies and narrow, flattened shapes that present the smallest possible light-reflecting surface to predators beneath them.

California gray whale, one of the baleen whales

6. Filter Feeding

In the sea, the animals that strain food from the water are called filter feeders. The filtering apparatus, which separates relatively few particles of food from comparatively large volumes of water, has caused species so equipped to evolve peculiar structures and habits.

SURVIVAL

Some animals, including many filter feeders, are sessile, able to move about only slowly or not at all. The barnacle is such a creature. Once settled on some stable surface, a barnacle never moves again. Barnacles are crustaceans, closely allied to the shrimps and crabs. The feathery appendages seen flicking in and out of a submerged barnacle are its feet, which have been modified over millions of years into food strainers. High tide brings billions of tiny animals and plants called plankton into the surf zone where barnacles live, and the duration of each tide is sufficient for a barnacle to filter enough plankton to sustain itself.

Sessile filter feeders rely heavily on tides and currents to bring them food, but some mollusks, like clams and oysters, are able to generate their own currents. These animals usually lie buried in mud or sand with two siphons extended. Water is pumped in through one siphon and expelled from the other. Food matter is removed from the water inside the animal's body. A clam can strain an astonishing twenty-five gallons of water a day in this manner.

Not all filter feeders are sessile. The herring is equipped with gill rakers that have been modified into sieves for trapping plankton. Bits of food are channeled from the gill rakers into the herring's esophagus, and from there to the stomach. The spaces in the sieve increase in size as the fish grows, allowing it to take progressively larger pieces of food while not retaining the smaller particles that sustained it through its early growth.

Ironically, the largest fish in the sea is not a predator but a filter feeder. The whale shark, a docile giant fifty feet long, feeds by straining water containing plankton through its gill rakers.

Baleen whales are also filter feeders. The giant blue whale, when feeding in rich antarctic waters, may strain three tons of planktonic crustaceans in a single day.

Sea raven camouflaged in seaweed

7. Camouflage

The ancient lobster trap almost crumbled as we lifted it into the boat. The float identifying its original owner had long ago been cut adrift by a storm or a boat propeller, and the lobsterman had chalked it off as a business loss.

When the trap was aboard we scraped away the seaweed and looked inside. Empty. Only the polished skulls of two sea robins, tied together once for bait, stared back. Just the same, we thought, we should open it and have a closer look.

There was a flurry of movement as the trap was tipped over. No one had seen the two large fish, so well camouflaged were they among the weeds and decaying lumber. Now they lay calmly on deck, formless and grotesque, with fleshy growths protruding from their bodies. Only their movement had given them away. They were sea ravens, bottomfish common in New England waters, and masters of disguise.

In cases of true camouflage an animal's external body form has been altered by evolution to resemble something else: a weedy rock, a lump of coral, perhaps a bit of flotsam. This often gives the creature a bizarre appearance. There are, however, tremendous advantages in looking more like a rock than a fish.

First of all, camouflage changes the fish's characteristic outline and makes blending with the background much easier. This is the protective aspect, enabling the animal to hide from its enemies. Second, camouflage provides a means of securing food. Prey species, disregarding what appears to be an inanimate object, venture near enough to be seized. Thus the strange and distinctive sea raven, formed as an aggregation of invisible atoms and molecules, must continue seeking invisibility to survive.

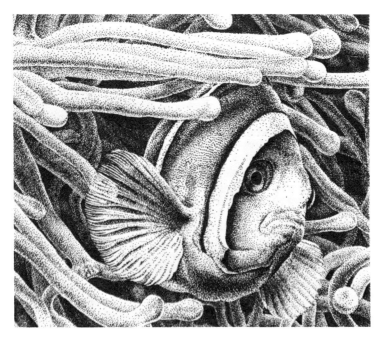

Anemonefish in tentacles of a sea anemone

8. Partnerships

Partnerships between two very different organisms have always intrigued mankind. In the fifth century B.C., Herodotus described how small birds entered the formidable jaws of basking crocodiles to feed on leeches and always emerged unharmed. Scientists now recognize this as mutualism, an arrangement beneficial to both parties: the reptile gets rid of unwanted leeches while the bird gains a meal.

In cases of commensalism one party benefits but the other is not harmed. Such is not true with parasitism, an arrangement in which one partner lives at the expense of the other. All three situations come under the general heading symbiosis, a Greek word that simply means living together.

The sea is so ancient that its inhabitants have had time to evolve complex interactions with one another. Perhaps best known is the case of sea anemones and a group of small damselfishes, the anemonefishes. The relationship is usually thought of as being commensal since only the fish gains an obvious benefit—protection from predators.

The sea anemone is a flowerlike creature that stings small animals to death before eating them. Special venom cells called nematocysts, located along the tentacles, perform the work.

The nature of the bizarre relationship between anemone and anemonefish puzzled scientists for years. How could this small fish live unharmed among the stinging tentacles of its powerful partner? In 1972 D. Schlichter, a German researcher, confirmed what many had suspected all along: the anemonefish is not immune to the sting; rather, by means of an acclimation ritual, it coats itself with the anemone's mucus, thereby masking its own presence and inhibiting the discharge of nematocysts.

Acclimation can be accomplished in an hour. When an anemone and an anemonefish are placed together in an aquarium, the fish's first response is to avoid its deadly antagonist. Soon it approaches the anemone cautiously. It may nip at the tentacles, or brush against them lightly. As the tempo of the ritual increases the fish bobs up and down above the anemone, touching the tentacles more and more frequently but never long enough to elicit fatal stings. Finally the tentacles relax. Acclimation is complete and the fish can burrow roughly across the surface of the anemone or push itself into the gaping mouth, all without being stung.

Immunity is not lasting. If fish and anemone are separated for a few days the ritual must be repeated; otherwise, the fish will be killed and eaten.

Jawfish demonstrating threat behavior

9. Territory

It was springtime on a Florida reef, and sunlight breaking through the surface played endlessly along the ocean floor. I slowly shifted into a cross-legged position, but the movement had no effect on the jawfish nearby. They continued to ignore me, even though my diving regulator bubbled intrusively.

The three-inch jawfish lives prairie-dog fashion in a burrow surrounded by the burrows of its fellows. As I watched, the two fish closest to my left flipper flared their gill covers threateningly at each other. The gesture finished, they retired to their burrows to await other challengers. The scene was strangely familiar.

In my yard at home a pair of robins had taken possession of a spruce tree the week before and were defending it against other robins. My tomcat patrolled the adjacent woods each night, seeking to exclude others of his kind. And my neighbor and I worked diligently to maintain the stone wall separating our properties. The defense of territory, it would seem, is almost universal among living things.

A territory is a defended area. Ownership of territory has definite advantages. It gives the defender enhanced energy: most animals fight fiercely when defending their homes. It also offers security from predation and a guaranteed food supply. In some gregarious species such as the jawfish, the territorial directive holds the group together. In many animals it acts as a means of genetic and birth control. Among male California sea lions, only those that succeed in gaining and defending a territory can breed. This limits the population, prevents the species from exceeding its food supply, and allows only the fittest to reproduce.

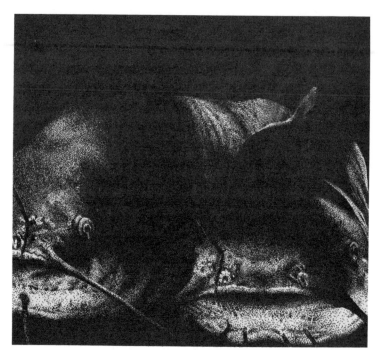

Compatible bullhead catfish

10. Chemical Language

Communication among animals is often chemical. Some of the substances produced for this purpose are called pheromones from the Greek *pherin*, to transfer, and *hormon*, to excite. Unlike true hormones, which are glandular secretions retained inside the body to regulate an animal's own physiology, pheromones are released into the environment where they profoundly influence others of the same species.

To succeed, a pheromone must be highly specific, so that it cannot be confused with other chemical stimuli, and it must be potent. (Pheromones are expelled in minuscule amounts. It is estimated that 0.01 microgram of pure gyplure, a pheromone excreted by female gypsy moths as a sex attractant, is theoretically sufficient to excite more than a billion males.)

Pheromones serve many purposes, but the most prevalent are to disperse the species, maintain social order, attract the opposite sex, and preserve genetic integrity.

Many animals excrete pheromones to disperse their own kind and prevent overcrowding of the habitat. Just how effective such a control can be was demonstrated with a species of East African snail. When the animals' pheromone was concentrated and added in twice the normal amount to a basin of water, all snails confined there died.

An alpha, or dominant, bullhead (a freshwater catfish native to the eastern United States and Canada) becomes aggressive toward a newcomer that poses a threat to its status. If the new fish is removed to another aquarium and some water from its container is added to the aquarium with the alpha fish, the latter emerges from its hiding place ready to battle the "intruder." Aggressive behavior continues until the offending chemical has dissipated. Had the new fish not presented a threat, the addition of water from its aquarium would not have elicited the same response. The same holds true even when the alpha fish is blind, indicating that bullheads recognize and remember individuals of their own kind by smell. Several different species of tidepool blennies, small fishes indigenous to California, look alike. To further complicate matters, courtship behavior among them is not highly varied, yet hybrids have never been found. This is because at mating time females excrete pheromones that are recognized only by males of their own species.

Among bullheads in an aquarium, a pheromone is released that depresses aggression and permits all residents to live in harmony. But if the olfactory tissue of one fish is destroyed it may persecute its tankmates unmercifully, apparently unable to recognize their chemical signals. Such antisocial behavior continues until the offending fish regrows its nose tissue.

Emaciated African lungfish newly emerged from mud cocoon

11. Air-Breathing

In the Silurian period, 415 million years ago, the earth was warm and swampy. Rivers spread over the land in the wet seasons, covering vast areas with a skin of water. But in the dry seasons, the water shrank back into stagnant pools and sinkholes where the sun fostered heavy growths of aquatic plants. In their frantic nighttime respiration, the plants monopolized the limited supply of oxygen, replacing it with the carbon dioxide which they gave off. Only the hardiest animals could survive in such a hostile environment.

Well before the beginning of the Devonian period, commonly called the age of fishes, 400 million years ago, a group of primitive vertebrates known as the placoderms had invaded the swamps. For thousands of years they hovered at the fringes, unable to leave the streams and rivers still left flowing when the other water courses dried up. At last the invasion started.

At first these ancient fishes used their gills to breathe the oxygen-rich surface water, but as time passed a few began to gulp the moist air just above the surface. Those that survived developed increased blood supplies around their gullets, a breakthrough that allowed them to take up oxygen at the air-water interface far more efficiently. In time, the gullet lengthened into a pouch, which increased the surface area for absorbing oxygen. Eventually the pouch extended back into the body cavity, forked around the internal organs, and became laden with blood vessels. This organ was a unique one among living animals of that time. It was a lung.

The appearance of a lung had two profound consequences. It made possible the evolution of land vertebrates, leading ultimately to mankind, and it also gave rise to the familiar swim bladder in modern fishes. The swim bladder greatly improved swimming efficiency and helped the fishes spread throughout the waters of the world. Today more than 99 percent of the bony fishes either have such a structure or evolved from ancestors that did.

Fishes with functional air-breathing lungs were once widespread, but today only six species of lungfishes remain, spread thinly across tropical Africa, Australia, and South America. Dependence on the lung varies greatly among species. The South American representative is totally dependent on air breathing and will drown if held underwater. The Australian lungfish seems not to need its lung at all, having perfectly functional gills as well. One of the African lungfishes survives the dry season in hard balls of clay, after the stagnant pools in which it is found evaporate. During this time it lives on accumulated body fat. With the first rains of the wet season the clay dissolves, and a thin but living lungfish emerges.

Mediterranean torpedo rays

12. Electricity

In all higher animals, several different types of tissue generate tiny electrical currents. The more than ten billion cells in the human brain each put out $2/100,000$ to $1/10,000$ volt. This is enough to measure with an electroencephalograph. But of all groups of animals, fishes produce the largest outputs of electricity. The electric catfish of Africa can produce nearly 350 volts. The electric ray of the North Atlantic generates 50 amperes and 60 volts. There are other species besides these, but most powerful of all is the electric eel, a native of the Amazon River drainage system in South America.

Electricity offers several advantages to a fish. For species able to generate enough power, electricity serves both as an offensive and a defensive weapon. Weaker species use it as a sensory guidance system, enabling them to navigate in muddy water and detect food and enemies. Finally, when utilized in certain ways, electricity can form a basis of communication among members of a species.

The electric organs in fishes evolved from muscle tissue. The special electricity-generating cells are called electroplaques. The electric eel has 6,000 to 10,000 electroplaques arranged in about seventy columns in each side of its body. The electric organs take up 40 percent of the animal's bulk. The cells are connected in series, which builds up the voltage. The parallel arrangement of the columns serves to increase the amperage. Electric eels grow to nearly ten feet in length, although captive specimens seldom exceed six feet. A six-foot specimen can produce 800 volts and one ampere.

The cells of the electric eel are insulated to keep the animal from electrocuting itself, but the eyes of adult specimens are normally cloudy, a condition probably resulting from their own discharges.

Electric fishes play an interesting role in the history of medicine. The torpedo, an electric ray found in the Mediterranean, was used in early treatments for gout. Scribonius Largus, a Roman physician of the first century A.D., kept his patients standing on a live torpedo "until the foot and leg up to the knee became numb." Ibm-Sidah, a Moslem physician who practiced medicine in the eleventh century, placed living electric catfishes on the brows of persons suffering from epileptic fits. Thus a catfish was the first instrument used in shock therapy.

Elephant seals resting on a pebble beach

13. Blubber, Fur, and Antifreeze

Marine mammals share with all other mammals the distinction of being warm-blooded, relying on the production of their own body heat instead of on the environment to keep warm. Land mammals inhabiting colder latitudes can burrow underground, dig into snowbanks, or take shelter in caves during severe weather. But the open sea, which is the only home of many marine mammals, is without dens or shelters of any kind. Escape from the cold is impossible.

The problems of retaining body heat in water are difficult, since water conducts heat roughly twenty-seven times more efficiently than still air of the same temperature. Thus a mammal that stays submerged is faced with the possibility of losing body heat continuously. Marine mammals have confronted the situation in two ways. First, most of them grow to a large size. Heat loss is to some extent a function of body weight versus surface area. Heavier animals have a smaller surface area per unit of body weight, a factor that cuts heat loss. Second, they have evolved specialized body forms and physiologic processes.

The cetaceans (whales, dolphins, and porpoises) are streamlined, with minimum surfaces exposed to the water. Only stubby flippers and flukes stick out. All cetaceans are hairless, or nearly so, relying on thick layers of fat, called blubber, to retain body heat. Blubber comprises up to 27 percent of the total body weight in the blue whale, and 23 percent and 21 percent in the fin and sei whales respectively. The blubber of the right whale may be twenty-eight inches thick in places and account for 45 percent of the animal's total weight.

The insulating qualities of blubber are further enhanced by a system of counter-current blood vessels. In other mammals the arteries and veins are arranged in nets that fan out through the underlying layers of tissue, but whale blubber is virtually nonvascular, permeated sparingly by bundles of arteries wrapped in veins. The arrangement substantially reduces heat loss from the blood. For example, colder venous blood returning to the heart from the flippers and flukes is warmed by arterial blood flowing in the opposite direction. Because it has already been cooled, the arterial blood has little heat left to waste by the time it reaches the outer appendages.

The pinnipeds (seals, sea lions, and walruses) are less aquatic than cetaceans. All pinnipeds breed on land and most return to land periodically to rest. Some, like the elephant seals, have hair but rely on blubber to keep warm. Fur seals, on the other hand, count heavily on their thick fur. Pinnipeds are able to shunt the flow of blood away from peripheral tissues when faced with situations demanding extra heat conservation. The result is that external tissues may be nearly as cold as the surrounding water, while temperatures deep within the body stay warm. But what about cold-blooded marine fishes?

This thought occurred to me one winter day while I was performing a chore all too familiar to those of us who live in the temperate zone—adding antifreeze to the radiator of my automobile. I followed the instructions on the container and ran the engine for several minutes to disperse the solution evenly throughout the cooling system. Next I checked the concentration in the water. Just right. The ethylene glycol was roughly 40 percent by volume. Very efficient. Or was it?

I stopped and considered. The blood serum of most marine fishes freezes at about 31 degrees Fahrenheit. By itself the fact is not remarkable. But it becomes unusual indeed when you consider that large expanses of the ocean are some two degrees colder still. How do fishes inhabiting the gloom of polar seas keep from freezing solid?

It is common knowledge that the freezing point of water can be lowered by adding salt. It would be logical to suspect that the blood of cold-water fishes might have a higher salt content than the surrounding seawater. But this is not the case. The blood of a marine fish is normally less salty than the sea. Then what was the answer? Antifreeze, naturally.

Many theories have been advanced to explain death by freezing. Cell damage caused by the formation of ice crystals is one possibility. This is a purely mechanical process, the same phenomenon that would have broken my car radiator apart.

But whereas freezing can cause death, mere cold does not. Fishes can live at temperatures below freezing provided their cells do not freeze. In such a supercooled state a fish will freeze and die if brought to the surface and experimentally "seeded" with ice crystals.

Cryoprotective, or antifreeze, compounds are present in the serums of many cold-water fishes. In some antarctic fishes these substances belong to the glycoprotein group and are responsible for the low freezing point of the blood serum. They operate by lowering the freezing point of water inside each cell in a fish's body without altering the melting point, and by forming layers over the surface of ice crystals, eliminating them as potential nucleation sites where additional ice crystals could grow. Efficiency? Glycoproteins comprise a mere 1 percent of the blood.

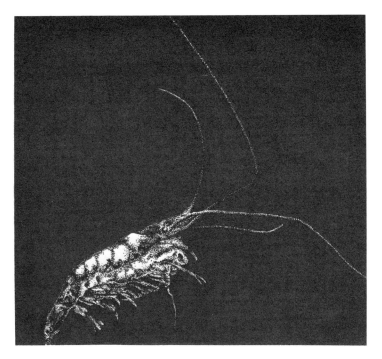

Luminescent shrimp

14. Biological Luminescence

The ability of some living organisms to glow with a cold green or blue light is called biological luminescence. The light in the water seen by Coleridge's ancient mariner as he lay becalmed near the equator was caused by heavy concentrations of dinoflagellates (primitive single-celled organisms) that luminesce when disturbed. Such ability is surprisingly common among the lower animals and plants. Many bacteria, fungi, sponges, corals, ctenophores (comb jellies), centipedes, and millipedes can luminesce. Probably the best-known luminescent organism is an insect, the firefly, and its larva, the glowworm. Among vertebrates, only certain fishes can generate and emit light, the technique having somehow been lost when amphibians emerged onto the land.

In 1887 a French biologist named Raphael Dubois, experimenting with a luminous boring clam, made a crude extract which he named luciferin after Lucifer, the "light-bearer." Another of his extracts was the enzyme luciferase.

But how is biological luminescence an aid to survival? And why is the phenomenon so widespread among living organisms? The survival value is often self-evident. One deep-sea squid discharges luminescent ink to erase its escape route. Several abyssal fishes have luminescent "fishing lures," fleshy appendages used to lure prey near enough to be captured. In many other species luminescence is of doubtful value.

As to the second question, one theory states that biological luminescence was simply a spinoff of normal chemical reactions evolved by primitive organisms to rid their tissues of oxygen. Early life forms functioned well in the absence of oxygen and poorly in its presence. The most efficient means of getting rid of oxygen is by reducing it chemically to water. Thus the ability to luminesce was perhaps retained by many evolving organisms, later to assume survival value in some species. This theory would explain why so many diverse animals and plants possess the remarkable ability to generate light.

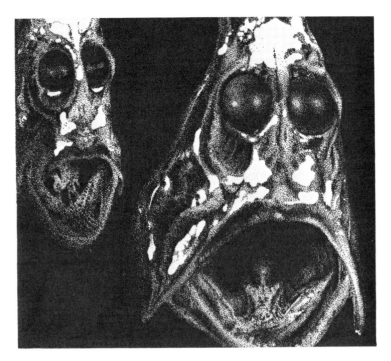

Abyssal hatchetfish

15. Abyss

Far below the ocean's tumbling waves and sunlight is another ocean, black and numbingly cold. This other ocean is saltier than the one we know. High saline water, squeezed out during the freezing of seawater near the poles, sinks and slides down the continental slopes into its domain. Currents are few, and the water lies torpid in black canyons, not reaching the surface again for centuries. There is great pressure, often seven or eight tons per square inch.

This other ocean is called the abyss, or deep sea. It is the largest single habitat on earth, and also the one we know least about. More than 80 percent of the earth's ocean is deeper than 10,000 feet; overall the sea averages 2.4 miles in depth, even including the shallow areas covering the continental shelves.

The abyss, though large, is a marginal environment for living things, and its inhabitants have adapted accordingly, taking on grotesque characteristics in the process. Abyssal fishes are noted for their dark color, usually deep sable or black. Crustaceans are often bright red. This would seem to attract predators, but red cannot exist as a color at such great depths. In the absence of currents and waves, even skeletal structures are hardly needed and have been greatly reduced in many deep-sea vertebrates.

Most abyssal animals are small, probably because the paltry food supply will not support an organism with a large biomass. Since meals may be infrequent, some fishes, notably the gulpers and pelican eels, are little more than swimming mouths.

Other adaptations are the evolution of luminous organs and enlarged eyes to see in the inky blackness. In several abyssal fishes, the light is actually produced by bacteria in the luminous organs. The bacteria need oxygen to glow. By stopping the flow of oxygen to its bacteria, a fish can control its glow.

Reproduction is another problem. How do abyssal organisms, never populous to start with, find one another? The deep-sea anglerfishes have hit on a bizarre solution. After spending its larval life in the plankton of the upper ocean, a maturing male anglerfish sinks into the abyss. If by chance he finds a female of his kind, his jaws latch onto her permanently. Eventually his digestive tract degenerates and his mouth becomes fused to her body. His eyes disappear, and he becomes a true parasite, living off the tissues of his mate.

Fingerfish with prominent lateral line

16. Extensions of Touch

Sonar, which in the terse abbreviations of our modern world is an acronym for *so*und *na*vigation *r*anging, has become vital in times of war and in peacetime oceanography. But whales, dolphins, and porpoises have been using a more sophisticated version of sonar for millions of years.

SURVIVAL

Sonar detects the presence and location of a submerged object. A series of sounds is emitted by an underwater transducer. The echoes bouncing back from the target are translated to show how far away the object is and its general acoustic-reflecting properties, or "target strength." Whale sonar is superior because it can perceive differences in the size, shape, and composition of an object. Blindfolded dolphins can tell a mackerel from a mullet, a feat as yet unmatched by the United States Navy. But how?

Perhaps color offers the best analogy. When white light, or light containing the full color spectrum, is shone on a green surface, only green is reflected; on a red surface, only red. A dolphin sends out a broad range of sound frequencies in rapid bursts. Some are absorbed by the target, but the rest are bounced back. Each target has distinct absorptive and reflective properties. To a dolphin, a mackerel "sounds" different from a mullet.

Cetaceans do not have vocal cords, yet they make a remarkable variety of sounds. Some are used in sonar and some in communication. Beneath the blowhole are air pockets branching away from the nasal passage. According to one theory, air forced into these pockets produces sound. The melon, or fatty forehead area, is thought by some to shape the sounds, much as an acoustic lens does. Returning echoes are picked up by the lower jaw and carried to the ear.

Long before whales appeared on the earth, fishes were using a different extension of touch. The horizontal stripe running along the side of a fish is its lateral line, an organ peculiar to fishes and a few amphibians. The lateral line is riddled with open nerve endings which are linked closely to the senses of touch and hearing. The system perceives different kinds of water movement, including such subtle factors as the flow disturbance set up by others of a fish's kind in the same school. From a survival standpoint, a prey species may "feel" different from a predator. The Germans call this sense *Ferntastsinn*, or distant touch.

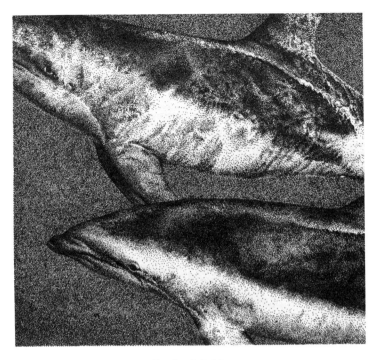

Dusky dolphins

17. Language and Intelligence

The engaging smile of the dolphin baffled men for centuries. What sort of intelligence did it conceal? The smile itself is fixed and can never be changed. But what about the complex "language" of dolphins, and their large and highly convoluted brain? Surely these things are signs of a native intellect perhaps equal to man's.

Sadly, for those who are warmed by myths, returns from the scientific community indicate that dolphins and dogs may be about equal in intelligence. The large brain is an organ adapted to processing sound and is probably not equipped to formulate complicated thoughts and abstract images.

Over the centuries, many myths have arisen about the "language" of dolphins. The fact is that, while dolphins can and do make an astonishing variety of sounds, they possess no language in the sense that we know. Scientists equipped with waterproof microphones have monitored dolphin sounds and projected them onto paper to create spectrograms, or sound pictures. The rise and fall of the noises emitted by a dolphin can then be compared with patterns of behavior. Dolphins do communicate vocally, but not in complex words and phrases.

Dolphins emit sounds of two types: pure-tone sounds, or whistles, and pulsed sounds. This second category can be further divided into those sounds used in sonar, and the various squawks, barks, bleeps, and grunts that many researchers believe are rudimentary expressions of emotions.

At first, scientists thought that whistle sounds formed the basis of a language. Recent findings indicate that 90 percent of the whistle sounds produced by a dolphin consist of a "signature whistle" used by the animal to identify itself. The spectrogram of any dolphin, then, would consist mainly of the animal's own special signature whistle. Not all scientists agree with this interpretation, but variations of the signature whistle may be what the dolphin actually uses to communicate. Instead of making other distinctly different sounds to express itself, the animal might alter the intensity and duration of its signature whistle. As an anthropomorphic example, if a dolphin's signature whistle can be translated to mean "I am Delphis," a frightened dolphin might express its fear by shortening the interval between bursts: thus, "IamDelphisIamDelphisIamDelphis." The same dolphin, when content, might repeat, "I . . . am . . . Delphis . . . I . . . am . . . Delphis . . . I . . . am . . . Delphis."

Male fiddler crab at entrance to its burrow

18. Biological Clocks

A solar day is the twenty-four-hour period it takes the earth to rotate once on its axis. But within the span of a lunar day, which is twenty-four hours and fifty-one minutes, there can be two tides, spaced evenly apart. Seashore organisms commonly predict tidal rhythms in advance. Many do this by means of a mysterious mechanism called a biological clock. A term sometimes used interchangeably is circadian rhythm, derived from the Latin words *circa* (about) and *dies* (day). It refers to a period roughly encompassing one day. When used in its proper context, it applies only to those biological rhythms that continue uninterrupted under constant conditions and is not meant to include activities that simply occur daily, influenced by such factors as daylight and darkness.

There is little doubt that the intrinsic timing mechanism of a biological clock exists at a cellular level. When the organs are removed from anesthetized hamsters, the intestine goes on contracting and the adrenal glands secrete their fluids as if nothing were wrong. An experiment in which eight free-living protozoan cells (*Paramecium*) were allowed to multiply for six days showed that the biological rhythms being examined had been passed along intact through the resultant 131,072 progeny.

It is still unclear whether biological clocks are escapement or nonescapement. In the former, the mechanism is thought to be set by the organism independently of outside influence. Proponents of the nonescapement hypothesis insist that some as yet undefined geophysical force signals the clock at intervals and "sets" it. Experimental evidence so far supports both hypotheses.

The fiddler crab is a conspicuous member of the intertidal zone of North American coastlines. When not involved in mating or threatening one another, fiddler crabs fan out across the mudflats at low tide to feed on microscopic algae.

SECRETS OF THE DEEP

Fiddlers only occasionally immerse themselves in the sea. The crabs enter their yard-long burrows just before high tide and plug the entrances with balls of mud. As the tide recedes they emerge to feed. Such rhythmic activities are innate, rather than learned or influenced by extraneous factors. The behavioral patterns remain fixed even under conditions of constant light or darkness, or when the animals are translocated to different time zones. Fiddlers indigenous to Massachusetts were flown to California, but even across a distance of three and a half time zones, their east coast tidal clocks remained unchanged for several days.

Greater shearwater in flight

19. Navigation

At south latitude 40 degrees, the Tristan da Cunha Islands are a scattering of rocks in the Atlantic Ocean, midway between South America and the tip of Africa. The greatest distance between any two of them is a scant thirty miles, but the group itself is fifteen hundred miles from the nearest land. This tiny outpost is the only breeding ground of a seabird called the greater shearwater.

After the breeding season greater shearwaters range into the northern hemisphere as far as north latitude 60 degrees. They move clockwise by the millions: north across the equator and into the western North Atlantic; past Bermuda to the jagged shores of New England; eastward across the Atlantic, brushing England and Spain and North Africa; then south to the islands, arriving there in November to breed again. Their journey is remarkable only to us. Many animals can determine both position and direction—in other words, navigate.

There is sufficient evidence to show that migrating birds can navigate by using celestial cues. Best known of these theories is the sun-arc hypothesis. It states that a bird can detect the movement of the sun across the sky and extrapolate enough of a completed arc to compare it with the home arc. Another related theory holds that birds can measure the rate of change in the sun's altitude.

The majority of migratory birds travel at night. Here again celestial cues are used, this time the position of the stars. During overcast nights when the stars cannot be seen, the birds often become confused and lost.

A previous theory suggested that birds might be guided by magnetic fields, but experimental evidence for this is lacking. The role of experience is minimal. Experienced birds that travel over land are known to make use of landmarks, an ability of little value to species migrating over wide expanses of ocean, or to young birds making their first migration. Hand-reared birds of several species have time and again proved their uncanny navigational prowess without ever having associated with older members of their kind. This demonstrates that the ability to navigate is innate, not learned.

How do birds know when they have migrated far enough? More experimentation is needed before this question can be answered conclusively, but to a traveling bird, the stars may again play a crucial role. Perhaps when the constellations have become sufficiently unfamiliar, the bird knows it has arrived at its destination.

Young seabird killed by an oil spill

20. Pollution

The tern pirouetted lightly above the waves, dipping and rising as the surface of the sea rose and fell. Although no enemies were in sight, still the little bird called out, *kee-urr, kee-urr*, in a never-changing pitch, as monotonous and rhythmic as the ocean's swell.

This bird was a male common tern, lately down from

the coast of Maine where he had bred that summer. Now the warm days were fewer and the signs of approaching winter had been his cue to start the annual migration south to the coast of Patagonia. The common tern makes the journey each year, across the equator and back again, although its wanderings are not so dramatic as those of its close relative, the Arctic tern. That little bird, not much bigger than a robin, annually makes a 22,000-mile round trip from the Arctic to the Antarctic.

Terns are the swallows among seabirds—small and airy with forked tails. Their bones are hollow, as are most bird bones, an evolutionary adaptation that reduces flying weight. This tern was very handsome: white underneath, with a black cap and black-tipped red bill. He was marvelously adapted to life at the interface between air and water; he could capture the tiny fishes that swam just beneath the surface or lightly touch down and bob corklike on the swells. But nowhere in his evolutionary heritage had he been programmed to cope with oil slicks. And so as he dived for a fish the sunlight caught his wings for the last time, and his wonderful adaptations were no longer of value.

After catching his fish he could not rise into the air. His white breast was smeared with oil, and his wings were sodden and smudged. The oil destroyed the water-resistant character of his plumage, and the cold ocean touched his skin for the first time.

For one day he fought against the oily sea until his breath rattled with pneumonia and his hollow bones filled with mucus. For one day he died slowly, and then at last, the small and tousled body sank from sight.

Part II

THE TROPIC ZONES

There is no hope—if not a coral
island slowly forming
to wait for birds to drop
the seeds will make it habitable

> —William Carlos Williams
> "The Descent of Winter"

21. Life in Tropic Seas

Winter in the city. The falling snow lasts for only an instant before turning to gray slush. All at once there are warm breezes, and I can feel myself sinking downward through clear blue water. Abruptly the illusion is erased by cold spray from a passing automobile, becoming once again an advertisement in the window of a travel agency. Surely the tropics are paradise, as the poster suggests. For humans, perhaps, but tropical marine creatures might express a different view, if they could.

Beneath the surface of tropic seas activity is frenetic, mainly because the water is too warm to support life comfortably. Heat accelerates the metabolism of cold-blooded creatures, shortening their life-spans. The near-absence of seasonal change has left tropical fishes and invertebrates ill equipped to tolerate fluctuations in environmental conditions, such as an unexpected cold spell. Finally, most tropical animals are smaller than species found in cooler waters. The warm water at first causes them to grow faster than their counterparts in mid-latitudes, then stunts their growth after sexual maturity.

Variety of life forms is the trademark of equatorial waters. Nearly 40 percent of all marine fishes live within the tropic zone, yet the tropics support fewer numbers of each species than either the temperate or polar regions. This fact has long puzzled scientists. Perhaps ecological niches in tropic seas are smaller because of stress and lowered tolerance to environmental change. Or are there simply more niches to start with, and of necessity they must be smaller?

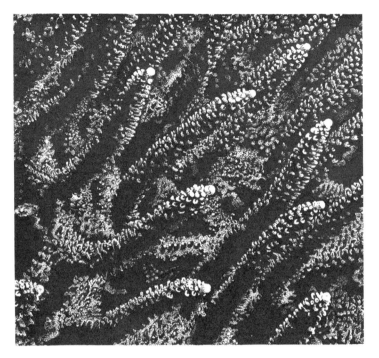

Staghorn coral

22. Shallow Reefs

The coral animal, called a polyp, is related to the sea anemone and sea jelly. Some corals are found in northern waters, but the spectacular reef-building types are limited to tropic seas where the water is free of turbidity, the temperature averages 70 degrees Fahrenheit the year round, and sunlight is intense. Such an area straddles the equator, forming a belt three thousand miles wide around the globe.

Coral reefs are of three types: fringing, barrier, and atoll. Fringing reefs develop adjacent to the shoreline and in shallow water. Barrier reefs are also found where the water is shallow but are separated from the shore by channels. An atoll is different, since it rises from deep water and forms its own island. Atolls are circular in shape with lagoons in the middle.

Despite their balmy surface temperatures and palm-laden islands, tropic seas are veritable deserts. The reefs teeming with fishes, the luxuriant growths of branching coral—both are misleading.

The clarity of the water and its color are the first indications that the quantity of life will be sparse. Water in the tropics is virtually free of suspended particles and dissolved organic matter. The deep blue color has two causes. First, seawater has its maximum light transmission in the blue region of the spectrum. Second, light entering the sea is selectively scattered upward by the water molecules. Since light with shorter wavelengths is scattered most, the result is a shift from greenish blue to just blue.

Particles, because they are larger than the wavelengths of light, would scatter the incident light back toward the surface, causing a return of the greenish-blue tone. In richer temperate seas additional color is caused by the yellowish tint of planktonic organisms and yellow compounds in solution. Thus the sparkling water of the tropics, though it encourages coral growth, discourages the abundance of life.

School of jacks patrolling a deep reef

23. Deep Reefs

Corals are limited to shallow seas. Below about two hundred feet the light is too dim for living plants to carry out photosynthesis. Corals themselves are animals, yet they culture one-celled algae called zooxanthellae within their tissues. This is a mutualistic arrangement whereby both corals and plants benefit.

Corals are carnivorous, feeding on tiny animals lumped together under the general heading of zooplankton. It was thought at one time that the zooxanthellae living in the tissues of coral polyps were used as an auxiliary food source. Now it is known that this is only partly true. Each individual algal cell secretes a sheath of carbohydrate about itself and this outer coating is what the polyps digest—not the cells themselves.

It is doubtful whether the massive structures built by the reef corals could have been formed without zooxanthellae, which utilize waste products of the coral (notably carbon dioxide and nitrogen and phosphorus compounds) and prevent them from accumulating. Perhaps just as important, photosynthetic activity by the zooxanthellae increases the deposition of calcium, a substance vital for the expansion of a reef.

Zooxanthellae cannot grow unless there is sunlight to support photosynthesis. The absence of these symbionts on the deep reefs is evident. The few corals near the two-hundred-foot level are gnarled and stunted, like the twisted pines of high alpine forests.

As sunlight is filtered through water its warm colors disappear. First to leave are the reds, at about thirty feet, soon followed by orange, and then yellow. The colors below about a hundred feet are paled and subdued. As a result, many creatures of the deep reefs substitute bright red pigment for black. To a diver viewing them in their natural surroundings, deep-reef animals appear as varying shades of gray. Since no reds can exist at these depths, their hidden brilliance goes undetected by predators.

The deep reef is home to many strange creatures: bushy black coral prized by jewelers, squirrelfishes with their large, liquid eyes, and other animals not often found in the sunlit upper reefs. It is a silent, haunting place, shrouded in eternal blue.

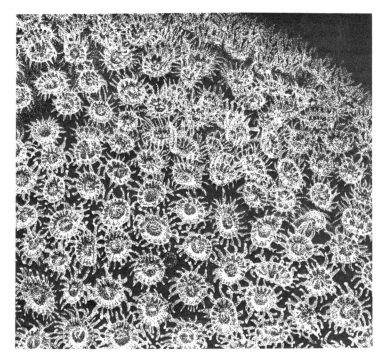

Expanded polyps of star coral

24. Reefs by Night

Different animals populate the shallow reef after dark. The milling schools of fish are gone, each individual having found a safe cleft deep within the coral. A few nocturnal fishes remain, but the open sections of the reef now belong to invertebrates. Sea urchins and sea stars stalk the dim canyons in search of food. Shrimps gather in great armies to feed on tiny bryozoans and other encrusting animals. Living corals resemble strange, delicate flowers as they extend to intercept the drifting dust of plankton.

Night-diving biologists have noted that color patterns in many fishes are drastically changed. Surgeonfishes and tangs are solid-colored during the day but striped at night. The nocturnal cardinalfish, brilliant red in daytime, turns ghostly pale when it emerges at dusk to feed. Besides changing color, some parrotfishes spin cocoons of mucus about themselves before retiring for the night. The purpose of the cocoon still puzzles scientists.

In some creatures the change in behavior and body form is so dramatic as to make them unrecognizable. A notable example is the basket sea star. During the day this relative of sea urchins and conventional sea stars rolls itself into a tight ball and hides in a coral crevice. At dusk it emerges and is transformed into a thing of nightmares. Its many arms extend and become a net to capture plankton and larval fishes. Each bit of food is grasped in a tiny fistlike projection, then passed "hand to hand" to the central mouth.

After sunset the spiny lobsters leave their burrows and walk about the reefs in search of whatever they can scavenge—dead animal material, algae, or living worms and mollusks. Lobsters have been shown to possess a keen homing instinct, and even though they may range for hundreds of feet, they unerringly return to their burrows before daylight.

Pipefish concealed in flowering turtle grass

25. Grassy Flats

In quiet lagoons and protected bays of tropic seas the bottom is often blanketed with turtle grass, so named because of the giant green sea turtles that graze upon it. In waters along the east coast of North America, turtle grass extends no farther north than the upper portion of the

THE TROPIC ZONES

Florida peninsula. Temperature affects the growth of turtle grass, as does water clarity. In turbid waters not enough light can penetrate for the plants to carry out normal photosynthesis.

Unlike the seaweeds, which are primitive algae, turtle grass is a true flowering plant, one of the few to be found in the sea. Its ancestors once grew on the land but gradually became aquatic. Turtle grass cannot survive unless completely submerged. Like other higher plants, it relies on cross pollination to propagate itself, but instead of being transported by insects and the wind, turtle-grass pollen is carried by water currents.

The dense growths of this foot-long plant provide a home for many specialized creatures, some of which cannot be found anywhere else. Myriad encrusting and attached animals and plants—more than a hundred kinds of algae alone—adhere to the blades. Many species of microscopic organisms grow to maturity and die without ever leaving the grass blades. Their presence attracts still larger creatures that feed upon them. Perhaps most specialized is the queen conch. This large marine snail feeds exclusively on algae attached to the turtle grass, but never on the actual grass blades.

Farther north along the west coast of Florida, the turtle-grass beds are more affected by seasonal changes. There is always a year-round abundance of sea urchins, sea stars, bay scallops, and hermit crabs. In spring the population swells to include vegetarian conchs and predatory mollusks such as the tulip snail. Many specialized fishes also appear: the spiny boxfish (a green-eyed puffer covered with sharp spines), cowfishes with their comical bovine faces, and seahorses and pipefishes, to name a few. These last two fishes have tube-shaped mouths for sucking in tiny crustaceans and other animals that abound within the miniature jungle of grass blades and runners.

Red mangrove branch with seedlings ready to drop

26. Mangrove Islands

The red mangrove is the "island builder" of tropic seas. Its floating seedlings are dispersed around the globe by ocean currents, germinating wherever the climate is mild and the water shallow enough for its roots to gain a foothold. The type of bottom is unimportant: red mangroves can even penetrate solid coral rock.

Tropic islands do not begin with mangroves. Conditions must first be favorable for the cigar-sized seedlings to be trapped and held long enough for their roots to take hold. The birth of a mangrove island results from a combination of factors. Perhaps debris or sand, driven by wind and storms, is trapped in shallow water among protruding branches of coral. The sand gradually deepens, raising the embryonic islet above sea level. Rain falling on the sand leaches away the salts, causing chemical reactions that transform unstable grains into a crude concrete. Now the islet can better withstand the force of the waves.

Eventually an itinerant mangrove seedling is trapped. The red mangrove is a halophyte, a plant able to live in seawater by maintaining the salt within its own tissues at a concentration greater than the surrounding ocean. Seedlings grow on the parent tree and drop into the water at maturity. A red mangrove seedling floats vertically with the end that will sprout roots always pointed down.

A red mangrove spreads outward as it grows. Bit by bit the stilted prop stems enlarge the island by trapping more debris. Near the center of the island, where the process initially started, leaf litter accumulates and is reduced to soil by scavenging amphipods (tiny crustaceans) and crabs. The soil turns into dry land ready to be colonized by conventional plants. These arrive as seeds—wind-borne, in bird droppings, or, in the case of palm nuts, with oceanic currents.

At maturity the island supports an astonishing variety of

animals: lizards and small mammals transported on floating logs and land crabs which have spent their larval stages in the sea. Birds roost in the trees, using them as a base from which to fish. Phosphate in their droppings enriches the soil but is also a major factor in the island's demise.

Phosphate accumulating on a living leaf burns and kills it. The island's decline begins when the trees die. Without a perimeter of living red mangroves to protect them, the tenuous soil and leaf litter give way. With the soil gone, waves break apart the concrete-like substratum and everything washes away.

Part III

THE TEMPERATE ZONES

The creeks overflow: a thousand rivulets run.
'Twixt the roots of the sod; the blades of the marsh-grass
    stir;
Passeth a hurrying sound of wings that westward whir;
Passeth, and all is still; and the currents cease to run;
And the sea and the marsh are one.

           —Sidney Lanier, "The Marshes of Glynn"

27. Life in Temperate Seas

Compared with the tropics, life in temperate zones is richer if not so diverse. The great abundance of living things is caused by a seasonal turnover of the water column that brings nutrients to the surface from bottom sediments. These nutrients—primarily nitrate and phosphate—trigger lush blooms of microscopic plants called phytoplankton, the first links in the food chain. Although there is greater illumination in the tropics than in temperate zones, the phytoplankton bloom near the equator is diminished because the more stable water temperatures prevent seasonal upwellings of nutrients.

Everything is varied at mid-latitudes. Tidal ranges are greater, creating enormous flushing action. This brings a surfeit of food to sessile organisms living in the intertidal zone and washes away waste products at an equal rate.

The proportion of large species is greatest in temperate seas. Animals live longer, though not so long as those inhabiting the polar regions. Also, temperate-zone animals show greater growth after attaining sexual maturity, whereas the opposite is true in tropic and polar areas.

In short, living conditions are most favorable in the mid-latitudes, declining in quality toward the poles and equator.

Whale catcher with dead fin whale in tow

28. Offshore Waters

The potential for a vast fishery lured the first Europeans to the shores of North America. In the sixteenth century the stocks of fishes found off the Grand Banks of Newfoundland, Georges Bank off Nantucket, and the inshore waters of New England seemed limitless. Captain John Smith, the English founder of the colony at Jamestown,

Virginia, stated after a trip to Maine in 1614, "He is a very bad fisher cannot kill in one day with his hooke and line, one, two or three hundred cods." Today those words have a hollow ring.

In 1974 the New England fisheries employed 12,645 full-time fishermen, a 30 percent decline from 1968. Landings of groundfish (cod, cusk, haddock, white hake, pollock, and ocean perch) in 1974 were 139.1 million pounds, representing a drop of 27 percent from 1968. The trend is typical of fisheries in all highly developed countries. Mankind is taking out more resources than the sea can put back.

As the size of the catch gets smaller, the better resource management so obviously needed is not sought. Instead, the fishing industry develops more efficient methods. Faster and bigger boats are built; increasingly sophisticated electronic equipment is developed to track the quarry relentlessly through murky seas and fog. Such improvements must be paid for. The inevitable result is even greater pressure to produce a bigger catch. The remaining stocks decline rapidly as more of their numbers are taken before they can breed and replace themselves. It is a vicious circle of shortsighted economics.

The great whales are a case in point. Many species are no longer found within their former ranges; others have been hard-pressed to survive even in the Antarctic, the remotest place on earth.

International regulation is the only solution; nationalism no longer has a place in worldwide management of marine resources. The whales continue to disappear; will the herring, the cod, and the mackerel go with them?

Snow geese feeding in a salt marsh

29. Salt Marshes

Estuaries and salt marshes are often confused by some people. An estuary is an area where a freshwater river or stream meets the sea. Normally it is baylike, a place in which fresh water and seawater mingle. A salt marsh is a low-lying swampy area behind a sandbar or barrier beach.

A salt marsh was once an estuary, but as tidal action increased the deposition of silt across its mouth, the body of water behind the silt became shallower.

A salt marsh begins as a mudflat. Channels form and meander down to the sea. Gradually the mudflat is colonized by grasses able to grow in brackish water. The grass collects sediment around its roots; in autumn it dies and falls over, forming a rich mulch through which next year's crop will germinate.

Grass forms the base of the salt-marsh food chain. Snails, amphipods, and other small scavengers shred the tough leaves and stems into fragments which quickly acquire a coating of bacteria and fungi. Microorganisms are rich in vitamins and proteins, and their presence increases the nutrient value of the grass particles upon which they feed. The result is that one year after dying, a patch of rotting marsh grass contains more than triple the protein it had when alive.

Eventually the grass is reduced to microscopic particles collectively called detritus. A healthy salt marsh produces five tons of detritus per acre each year. Detritus nurtures the minute crustaceans and mollusks, which in turn serve as food for larval fishes that are hatched in the salt marshes or drift in with the tides to spend their early lives there.

Salt marshes are among North America's most valuable natural assets. They once stretched for 3,700 miles along the Atlantic and Gulf coasts, forming a buffer zone against the ravages of ocean waves. Now many are gone, either dredged and filled for housing and commercial developments, or slowly strangled by dikes and dams.

SECRETS OF THE DEEP

The beneficial effects of extensive salt marsh areas are far-reaching. Unused nutrients concentrated in a marsh flow into estuaries to nourish phytoplankton, often doubling the quantity of life within the estuary.

Salt marshes are the most productive acreage on earth. A healthy salt marsh produces ten tons of organic matter per acre per year, whereas the best farmland in the midwestern United States yields only one and a half tons of wheat per acre per year, including leaves and stems.

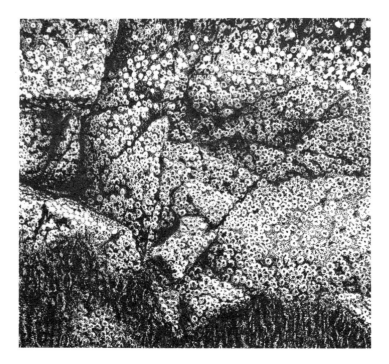

Zones of life on a rocky shore

30. Intertidal Zone

New England's rocky shore is divided into five different horizontal zones of plant and animal life. Within each zone live species that seldom populate the others. Starting from dry land and working seaward, it soon becomes evident that what distinguishes one area from the rest is the amount of water it receives in each twenty-four-hour period. The best time to study intertidal zonation is at low tide. The uppermost rocks, reached only occasionally by splashes from the highest waves, are called the lichen zone because of the primitive grayish-green plants found there. Lichens are hardy, and the haphazard drift of salt spray is enough to keep them alive.

The next area down is dominated by blue-green algae. This location is easily distinguished from the others by its black line—at low tide it looks like a dirty bathtub ring. Blue-green algae can also withstand drying out during low tide, although their habitat receives considerably more water than the one above it. Both the lichen and algae zones are above the high-tide mark. Insects from the land, small crustaceans called isopods, and crabs on their way to being terrestrial all scavenge together among the blue-green algae for bits of organic matter.

The third region is splashed by each high tide, though seldom submerged completely. The principal inhabitant is a barnacle, *Chthamalus*, that over millions of years has evolved the ability to withstand prolonged exposure to the air.

The fourth zone is also dominated by a barnacle, but of a species called *Balanus* that requires more moisture than *Chthamalus*. While the two live side by side, they seldom intermingle.

Fifth from the top is the rockweed zone, characterized by limp strands of rubbery rockweed able to tolerate brief periods out of water. Below the low-tide mark strands of kelp stretch seaward from the rocks.

The organisms that most typify intertidal zonation are the periwinkles, small snails that browse on algae. Periwinkles are fascinating subjects in the study of evolution because some species are in the process of leaving the ocean to take up life on land. The smooth periwinkle is still bound to the sea. Its forays out of the water are brief and uncertain. The common periwinkle, found in the middle zones, is more advanced in its ability to live on land, whereas the rough periwinkle can live above the high-tide line since its gill functions like a lung. All three species evolved from a common ancestor.

Spatterdock growing in a quiet pond

31. Freshwater Ponds

Ponds are small standing bodies of water with uniform temperature throughout. Each pond is unique, having its own peculiar balance of animals, plants, and microscopic life. The water in a pond may be fresh, salt, or a mixture of both, but most people think of a pond as having fresh water. Dragonflies hover in the still summer air, frogs belch from the banks, and fat bass and bluegills rise impatiently to worms dangled from cane poles.

The life-span of a freshwater pond is short when compared to more enduring features of the land. Before a mountain range can be created and worn away, millenniums must pass. But a pond may be born, reach maturity, and finally die of old age within a man's lifetime.

A new pond gouged from the earth by a bulldozer is barren of life, but within a few days living things become established in it. Water birds bring the eggs of aquatic insects and plants on their feet. Small trickling streams form a pipeline for fishes and their eggs, other insects and plants, and crayfish. On wet spring nights frogs, salamanders, and turtles travel to the new habitat by overland routes. Soon there is a thriving community.

As the pond matures, the life within its boundaries is sorted and categorized. Niches are carved out and occupied by specific animals and plants to the exclusion of others. Cattails grow only in the shallow portions along the shore; water lilies, fanwort, and others prefer more complete submersion. Air-breathing frogs live in the weed-filled shallows, but bass and other fishes are found in deeper water near the center. On close inspection it is evident that a pond is not a disorganized jumble of life but a highly structured community of animals and plants.

As the pond grows older its nutrient composition increases, a process called eutrophication. The profusion of life can only bring an equal profusion of death. Plants and animals die and decompose, releasing their nutrients back into the water. The ever-increasing nutrient material triggers still more plant growth. In old age the pond is choked with weeds and sediment and its quantity of animal life declines. The shoreline creeps ever closer to the center, first forming a bog, and in the end a wet spot in a meadow.

Leftovers of a seal rendering works

32. Rocky Coastlines

Rocky coastlines are fragile outposts of life, having proved especially vulnerable to human greed. Many coastal birds and mammals are now extinct, or nearly so. The heedless decimation of northern fur seal and sea otter populations along the American west coast is proof of a once insatiable fur industry. The bleached bones of thousands of elephant seals along the rim of the southern oceans are grim reminders of the living animals that were clubbed senseless, their flesh rendered into oil. In the Western Hemisphere no case of wanton slaughter is more dramatic than that of the great auk.

The great auk was a large fish-eating bird standing about three feet tall. At one time its range extended along the European coast from northern Norway to the Baltic Sea. It lived around the edges of the North Sea and inhabited all the shores of England, Ireland, Iceland, and Greenland. In North America, the great birds ranged from Labrador and Newfoundland south to Florida. The term penguin, which is derived from the Welsh *pen* (head) and *gwyn* (white), referred originally to the great auk. When "penguins" were later discovered, the name was applied to them too. But the birds we know as penguins all live south of the equator.

Great auks were once present in enormous numbers throughout their range, but they were flightless and easy to kill with clubs. (Early mariners in North America killed them for food.) They also bred slowly, producing only one egg a year, and the two factors drove them quickly to extinction. Two hundred years after the colonization of America, no great auks could be found there.

The great auk also disappeared from Europe, though no one knows why, since it was not hunted extensively for food. The last European great auk was caught in the harbor at Kiel on the Baltic Sea in 1790. By 1800 the bird's world range had been reduced to the Icelandic coast.

Early in the nineteenth century it was evident that great auks were disappearing, and museums started offering large sums of money for their skins. No thought was given to preserving them as living wild creatures. At daybreak on June 3, 1844, two Icelandic fishermen named Jon Brandsson and Sigourour Islefsson killed the last two great auks on Eldy Island off the coast of Iceland. A companion, Ketil Kentilsson, smashed the pair's single egg. At that moment the great auk became extinct.

Sea otter framed in kelp

33. Kelp Beds

The dominant marine plants are microscopic single-celled algae that drift in the upper sea, proliferating and dying off as the seasons change. But around the fringes of the land grow seaweeds. With few exceptions they are algae.

(98)

Most conspicuous are the green, red, and brown algal groups, some species of which grow to enormous size.

Several of the brown algae, called kelp, are the longest plants on earth, often attaining lengths of several hundred feet. When sunlight and other conditions are ideal, some forms of giant kelp grow at a rate of two feet per day.

North of the equator, beds of giant kelp are found only in the chilly arc of the North Pacific. Starting from southern California, giant kelp can be found just outside the surf zone along the coastlines of Oregon, Washington, British Columbia, and Alaska. After following the Aleutians, it ranges southward through the western North Pacific to the islands of Japan.

California kelp beds are dense forests within which thrive a multitude of life-forms. The abalone, a vegetarian mollusk, crawls slowly along the ocean floor feeding on smaller species of algae. Sea urchins, also vegetarians, sometimes graze on the kelp itself. Spiny lobsters find refuge among the holdfasts at the bottom, whereas many kinds of fishes—garibaldi, kelp bass, and kelp greenling—find shelter and food among the fronds. Attached to the surfaces of the fronds are countless encrusting organisms: coralline algae, barnacles, hydroids and sea anemones, and sea squirts.

The top of the forest canopy is the domain of one of the sea's most beguiling creatures, the sea otter. Sea otters are found mainly in kelp beds. They spend their lives among the fronds at the surface, resting, sleeping, even giving birth there. When hungry, a sea otter dives to the bottom and picks up a sea urchin, abalone, crab, or some other invertebrate and brings it to the surface to eat. If the catch has a hard shell, the otter also brings along a flat rock. It then rolls over on its back at the surface, places the rock on its chest, and cracks open the shell by pounding it on the rock.

Towering wave in mid-ocean

34. The Open Ocean

As winter begins to wane the open sea explodes suddenly
with microscopic plants. The initial bloom is modest, a
mere harbinger of the phytoplankton hordes to follow.
The sea in winter has been restless. Violent storms have

ripped at it relentlessly, and cold winds have chilled its surface waters until they grew dense and sank into warmer layers below. The combination of storms and cold established a pattern of vertical mixing and an upwelling of nutrients from the bottom sediments.

As the days grow longer and warmer and the light becomes more intense, the tiny plants multiply. Foremost among them are the diatoms, which are able to convert the sun's energy into living tissue. Each diatom fuels itself on sunlight and the rich nutrients brought to the surface in winter. It then divides, splits its silica case, and forms two new diatoms. These too divide and divide again until the surface of the sea is stained brown with them.

At their most numerous, the diatoms are attacked by copepods, tiny crustaceans which have perhaps the largest population of any animal on earth. With the copepods come mysids and amphipods and uncountable other crablike creatures that move with armored legs through the liquid swells. The copepods graze on diatoms and reproduce until the sea is filled with their legions.

Now transparent arrow worms appear from nowhere to attack the copepods and their kin. The swallowed copepods line up like billiard balls within the invisible bodies of their captors.

As spring lengthens, the herring come to feed on plankton. The ctenophores, drifting blobs of jelly, mass at the surface to trap larval herring and the young of barnacles and crabs. And as the adult herring churn in profusion through this teeming broth, they are slaughtered by bluefish arriving from the south, and by sharks from inside the continental shelf.

By summer it is over. The bodies of the dead and dying sink to the bottom, returning their nutrients to the sediments. Next spring will bring another resurrection.

Sanderlings feeding in the surf

35. Ocean Beaches

The common sand of ocean beaches is composed of quartz and feldspar and results from the weathering of gneiss and granite rock. Sand grains at a beach are the same size because the rate of settlement varies according to mass. Larger, heavier grains settle out first near the high-tide line. Since wave action diminishes seaward, the finer grains settle last. Sand grains are rarely worn down into smaller particles. A thin coating of water about each submerged grain acts as a bumper to protect it from the abrasive action of adjacent grains.

If one word could describe conditions at an ocean beach, it would be impermanence. Tides, winds, and currents combine to make this the most unstable of all marine habitats. Dry grains above the berm (the high-tide line) are blown about by the wind; those exposed to tides and currents are moved along by littoral drift. Even the size of a beach varies with the seasons, being wider in summer when sand is moved in from offshore bars. In winter the berm shrinks; storm waves keep the sand in suspension longer, resulting in more of it being deposited on bars.

A dominant though inconspicuous creature of the berm is the quarter-inch beach flea, one of the few crustaceans able to live for extended periods out of seawater. The beach flea is an important scavenger of the intertidal zone, feeding on organic matter left by the falling tide. Beach fleas spend the day in individual burrows above the high-tide line, emerging to feed at evening.

Another common inhabitant of the berm is the ghost crab. Like the beach flea, ghost crabs have mastered the problems of living away from the ocean. A ghost crab carries its own water supply inside its gill chambers and enters the sea only occasionally to renew it.

Intertidal residents of an ocean beach have adapted to life in the precarious shifting sand. Clams and the moon snails that prey on them are expert burrowers. Even the sea stars are different: some species do not have suction cups on their tube feet, having evolved in an environment without stable surfaces to cling to.

Dune wolf spider

36. Sand Dunes

Dunes form behind wide ocean beaches where the sand is not wetted by the tides. Wet sand cannot be blown about, and wind is the most important factor in dune formation.

Only a few species of hardy plants are able to grow on

the upper beach. Instability of surface, a deficiency of nutrients, and a harsh, biting dryness make it an undesirable habitat. To counter the effects of being buried much of the time, dune plants have evolved mechanisms that allow life to continue. One technique commonly employed by dune grasses (the key plants in dune formation) is to send out underground stems called rhizomes that erupt at different locations when the main stem is buried. Still other species continue to grow vertically, somehow managing to poke through the tops of even the tallest dunes.

A dune is born when a piece of driftwood or a tangle of seaweed traps the wind-blown sand and creates a small mound. If the mound can be colonized by beach grass with its hidden net of stems, a dune will grow, fed and nurtured by new sand brought by the wind. If enough rain falls to moisten the plants, and if the berm is sufficiently wide to relinquish a continuous supply of new sand grains with their thin coating of nutrients, other plants will colonize the area: bayberry, beach plum, and the salt-spray rose, to name a few common ones.

Dunes cannot be built without the aid of wind. Saltation is the process by which sand is carried along in the wind. The first sand grains set in motion must eventually fall back to earth. Upon hitting other loose sand grains at the surface of the beach they bounce back up, but not before setting these in motion also. Looked at closely, a dune seems to lean into the wind. Aeronautical engineers know that the laminar flow over an airfoil crowds the streamlines and causes the air to gain speed near the point of the curvature. The same phenomenon takes place in dune formation and explains why a sand dune leaning seaward has a shape reminiscent of an airplane wing.

If conditions on a dune remain stable, thickets spring up behind it. Raccoons, skunks, and foxes use the thickets as a permanent base from which to forage in the dunes for food. White-footed mice scratch for seeds in the thin loam and in turn are hunted by circling hawks. And wolf spiders stalk insects through the dry scatter of last year's grass.

Part IV

THE POLAR ZONES

> . . . but there is something
> in the contours of rock
> that always remembers
> both seasons,
> the ice and the sun.
>
> —Loren Eiseley,
> "Men Have Chosen the Ice"

37. Life in Polar Seas

The climate of the Arctic is oceanic and therefore milder than that of the Antarctic, but the curvature of the earth causes sunlight to shine at an ever-flattening angle toward both poles. Thus the ends of the earth are places of intense cold where for half the year the sun turns monotonously overhead, never falling below the horizon, and daylight is perpetual. When at last the sun sets, this phase, too, lasts for half a year and the poles are in a state of darkness seemingly without end. This drama of long days and nights is symbolic of the extremes to which nature has gone in these frigid, hostile worlds.

In the tropics the terrestrial fauna and flora are lush and the oceans virtual deserts. This condition is reversed at the poles. There the land is bleak and empty, whereas the seas are rich and infinitely more varied. In many other ways the tropic and polar zones are strangely similar, resembling each other more closely in some respects than either resembles adjacent temperate seas. Both regions are noted for their reduced tidal ranges and paucity of intertidal life, the low salinity of their oceans, and animals that follow summer spawning cycles and grow little after sexual maturity. In short, the poles, like the tropics, are places of intense environmental stress.

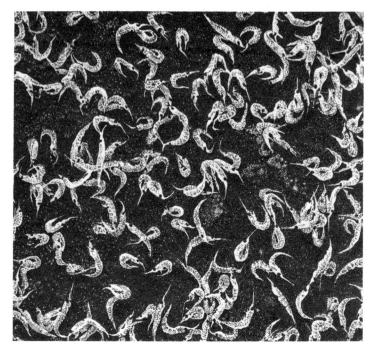

Krill drying on an antarctic beach

38. The Polar Seas

The oceans of the world average a chilly 39 degrees
Fahrenheit. Thus most marine creatures are accustomed to
cold. Among the factors affecting the biosphere, light and
temperature dominate, holding sway over such critical
functions as breeding, migration, and orientation. But no-

(114)

where are their effects so dramatic as in polar seas, where temperature is relatively constant but the cycles of daylight are measurable in months instead of hours.

During the summer in the Antarctic and in the Arctic six months later, food production is intense. In winter all production stops; living organisms idle or become quiescent, waiting for the long day to break again. In summer the constant light is augmented by an upwelling of nutrients so rich that even at season's end the stores are not depleted. In the Antarctic Ocean, which is circumpolar and reaches from the edge of the continent for two thousand miles in all directions, the turnover in the water column is brought about by the outflow of melted ice, by internal currents, and by a natural inversion created when the fresh water of the melting ice meets the warmer salty water farther north.

Food chains at the poles are the shortest in the sea. Primary production still begins with a bloom of phytoplankton (the primary producers), which in turn is fed upon by zooplankton. But there the similarity ends because the crustaceans comprising the secondary producers are large and numerous, far more so than their counterparts in the tropic or temperate zones. In the Antarctic these organisms, called krill by whalers, may grow to lengths of two and a half inches. Their counterparts in the Arctic, which are copepods, are much smaller. In both cases the secondary producers are fed upon directly by creatures at the apex of the food chain, instead of by slightly larger intermediate animals.

Baleen whales browse through vast stretches of ocean made rich by the russet bodies of krill or copepods. Seabirds can forgo a steady diet of fishes because the teeming billions of crustaceans in the upper nine meters of the sea are large enough to sustain them. In the Antarctic there is even a seal, called the crabeater seal, with teeth especially adapted for straining krill from the icy waters which surround the continent.

Emperor penguins "tobogganing" over the ice

39. Sea Ice of the Antarctic

At the end of austral winter the sea ice (ice formed from freezing seawater) breaks up and is carried north, forming a circle of low icebergs about the continent. This heaving, grinding barrier is known to mariners as pack ice. Its further progress is stopped at about 65 degrees south latitude by prevailing westerly winds.

The freshwater ice brought by glaciers from the interior of the continent also breaks apart in spring and drifts north, but the great mass of these freshwater bergs carries them beyond the pack ice to perish in the warmer waters of the lower latitudes.

Freshwater icebergs are too tall for seals and flightless penguins to inhabit. These animals use the lower bergs of sea ice instead, and two species—the emperor penguin and Weddell seal—are creatures of the sea ice throughout their lives.

Of all seabirds found in the Antarctic, the emperor penguin is the only species that breeds on sea ice frozen to the edges of the continent. Other birds of the region breed on outlying islands. Emperor penguin females lay a single egg in May, whereupon the male of the pair takes it on his feet and tucks it under a flap of skin. The males then crowd together in "huddles" to retain body heat in an environment that besides being dark may drop to –80 degrees Fahrenheit. In deepening winter the females return to the sea to feed over an ever-lengthening shelf of sea ice, often walking sixty miles or more before reaching open water. For two months the males stand in the bleak antarctic landscape, incubating the eggs and living off their body fat. At last their mates return and take the eggs on their own feet, freeing the males to feed. The eggs hatch soon afterward and the females feed the chicks food regurgitated from their crops. From then on, males and females alternate on feeding trips across the ice. By winter's end the maturing chicks are left in huddles of their own to be carried north on the icebergs starting to form under them.

Of four species of seals found in the Antarctic, the Weddell seal is best known. The others (leopard, crab-eater, and Ross) are confined chiefly to the ring of pack ice, but the Weddell seal, like the emperor penguin, winters on the sea ice itself. Its prowess in this hazardous world is incredible: adult seals can dive to depths of nearly two thousand feet or swim several miles on a single breath, staying down for an hour or more. In the dead of winter the animals spend most of their time underwater, where temperatures, though substantially below the freezing point of fresh water, are still warmer than the atmosphere. The seals maintain breathing holes by means of serrated teeth especially adapted to gnawing away at the ice.

Wandering albatross in flight

40. The Sea and the Air

It is unfortunate that the bird named *Diomedea* by the Swedish botanist Linnaeus in the eighteenth century has come to be called by English-speaking sailors "goney," or "gooney," terms alluding to simple-mindedness. Linnaeus was closer to the truth: Diomedes was the Homeric hero whose companions were changed into birds. Albatross is the name most of us recognize, a corruption of the Spanish *alcatraz*, meaning pelican.

But no name assigned to this magnificent creature can belittle its remarkable achievement: of all the birds it stands alone as true master of the air. The air, in fact, is so much its domain that several species of albatross rarely see land at all and never touch down on it except to breed.

Albatrosses belong to the order Procellariiformes, all members being characterized by nostrils in the form of tubes on top of the bill. Most famous of the group is the wandering albatross, a bird with a wingspan exceeding eleven feet, greatest in the avian kingdom. The wings extend outward from a body that may measure some fifty inches from beak to tail but weighs a mere eighteen pounds, further proof of the bird's commitment to life in the air.

The wandering albatross is confined to the southern hemisphere; more particularly to the immense southern ocean (sometimes called the Antarctic Ocean) where the Atlantic, Pacific, and Indian oceans converge. The weather there is the most tempestuous on earth because no land mass interrupts the wind as it circles the globe. There is nothing except the sea and the air.

Having evolved in this desolate outpost of life, the wandering albatross has succeeded with élan. Its young develop leisurely, not leaving the nest for nearly a year. During much of the time the fledglings are left unguarded by their parents, who need not fear predators on the lonely oceanic islands where they breed. Wandering albatrosses do not breed for their first ten years, and then a pair produces only a single egg every other year.

From the time a young bird learns to fly until its death (which may not be for fifty years), a wandering albatross goes where the wind takes it. In the southern ocean the prevailing wind is westerly, so the great birds glide eastward around the earth, circumnavigating the globe endlessly. Thus they do not fight the wind but rather ride with it, knowing that eventually all familiar places will pass beneath them again.

Epilogue

The future of marine life is threatened by the increasing contamination of the oceans. It may not be an exaggeration to predict a desolate shoreline for the twenty-first century.

The scene is a river bank. The late afternoon sun is red and swollen as an irritated eye. An old man sitting on the bank fishing is spotted by a boy.

"What are you doing, old man?"

"Fishing," the old man replies without looking up.

"What will you catch?"

"Nothing."

"Then why do you fish?"

"To catch nothing. To pass time doing nothing."

The boy scrambles down the bank and stands beside the old man. "My grandfather told me that people fished here. But that was long ago. Were there many fish?"

"Many fish. I used to catch and eat them fresh from the river."

"Did they look like fish when you ate them? Weren't they ground to powder and detoxified first?"

"There were no poisons in the water."

The boy sits down on the muddy bank where nothing will grow. "What made the river die, old man?"

"Greed. Politicians who sold out to industry. Thoughtlessness."

"I found a bird's nest today. There were three babies in it."

"What sort of birds were they?"

The boy lies back and looks at the sky. "I don't know. They will not live. Each had four legs."

"When I was a boy a bird with four legs was an oddity. Now such a grotesque creature is commonplace." The old man reels in his line and glances up at the sky. The line is caked with oil from the river. It has no hook at the end.

"Did you fish in the ocean when you were a boy?"

"Yes, often."

"What was it like?"

"Clear and blue and the salt spray on your lips tasted clean and bitter. Now the ocean is fetid and oily. The ocean is dead. Soon the winds will turn dry as dust and starlight will burn us. Men will go mad." The old man spits on the ground.

The boy shivers. He thinks the sky looks redder than before. "What will happen to us now?"

GLOSSARY

SELECTED BIBLIOGRAPHY

INDEX

Glossary

ABYSS The deep ocean beyond the edges of the continental shelves where the depth is 8,000 to 19,000 feet and the temperature is never warmer than 39 degrees Fahrenheit.
ADAPTATION Modifications in behavior or body form that help an organism to survive.
ALPHA FISH The dominant fish in a group.
AMPHIPOD Crustacean of the Order Amphipoda in which the body is compressed from side to side (laterally) and arched. Most amphipods are marine; some can survive out of water for extended periods.
ATOLL Coral reef, roughly circular in shape, with a lagoon in the middle. (See also Barrier reef and Fringing reef.)

BARRIER REEF Coral reef separated from the shore by a channel. (See also Atoll and Fringing reef.)
BASAL DISC The flat, bottom part of a sea anemone. The basal disc is an adaptation for securing the anemone to a stable surface.
BERM Ridge of sand on a beach that limits the swash of waves at high tide.
BIOLOGICAL CLOCK Internal timing mechanism in living organisms enabling them to predict regular natural events, such as tidal changes, and to alter behavioral or physiologic patterns and processes accordingly. True biological clocks continue to function normally under constant conditions. (See also Escapement hypothesis and Nonescapement hypothesis.)
BIOLOGICAL LUMINESCENCE Green or blue light self-generated by many lower animals and plants. (See also Luciferase and Luciferin.)
BLUBBER Layer of fatty tissue encapsulating the bodies of most marine mammals. Blubber provides thermal insulation.

SECRETS OF THE DEEP

BYSSUS THREADS Threads manufactured in some mollusks by the underneath portion of the foot which secure the animal to a stable surface.

CETACEAN An animal of the Order Cetacea, a group of marine mammals including the whales, dolphins, and porpoises.

CIRCADIAN RHYTHM See Biological clock.

COMB JELLY Animal of the Phylum Ctenophora. All species are translucent marine forms, characterized by eight rows of comblike cilia used for locomotion; hence the common name, "comb jelly."

COMMENSALISM A symbiotic arrangement between two very different organisms in which both parties benefit. (See also Mutualism, Parasitism, and Symbiosis.)

COPEPOD Small crustacean of the Order Copepoda, having an elongated, jointed body broader at the front end. Copepods are highly variable in habitat and life-style, with some forms being parasitic. Most, however, are free-living. Copepods are vital secondary producers in many oceanic food chains.

CORAL POLYP An individual in a colony of living coral animals. Each polyp resembles a small sea anemone, having a central mouth surrounded by stinging tentacles. Individual polyps are fastened to cups of calcium carbonate, which the animals secrete about themselves. The bleached corals sold in curio shops are aggregations of these empty cups.

CRABEATER SEAL One of four antarctic seals (the others being the Ross, leopard, and Weddell seals). Crabeaters are swift swimmers and perhaps the most numerous seals on earth. Their teeth are adapted for straining krill from the water, hence the name "crabeater," although "krill-eater" would be more accurate.

CRYOPROTECTIVE COMPOUND Any compound, such as automobile antifreeze, that lowers the freezing point of another substance and keeps it from freezing. Several polar and temperate fishes produce cryoprotective, or "antifreeze," compounds to prevent ice crystals from forming in their blood.

CTENOPHORE See Comb jelly.

DETRITUS Disintegrated organic matter. In salt marshes, detritus forms from decaying marsh grass, often serving as a substitute for phytoplankton at the base of salt-marsh food chains. Bits of detritus become covered with bacteria and fungi, which, in turn, increase the nutrient value of the particles for organisms next up in the food chain.

DIATOM Primitive single-celled plants of the Class Bacillariophyceae. Diatoms are widely distributed in fresh, brackish, and seawater. Many can move about under their own power. Diatoms differ from other algae by having a cell wall composed of silicon. Reproduction is normally by cell division.

ELECTROPLAQUES The special electricity-generating cells in electric fishes.

ESCAPEMENT HYPOTHESIS Hypothesis supporting the concept that biological clocks in living organisms are "set" by the organisms themselves, and that the resulting rhythms function independently of outside influence. (See also Biological clock and Nonescapement hypothesis.)

ESTUARY Transition zone between a river and the sea, where fresh and seawater mingle with every change of the tides. Estuaries are highly productive, serving as nurseries for larval fishes and as habitats for many valuable shellfishes and crustaceans. (See also Salt marsh.)

EUTROPHICATION Aging process of a lake or pond in which plant life becomes overproductive because of an increase in the abundance of plant nutrients.

FILTER FEEDER An animal that strains its food from the water.

FRINGING REEF Coral reef growing close to shore. (See also Atoll and Barrier reef.)

FROND The body, or main part, of an algal plant.

GLYCOPROTEINS Cryoprotective compounds in the blood of some fishes that function as "antifreeze" by preventing the formation of ice crystals. (See also Cryoprotective compound.)

GREAT AUK Largest member of the Family Alcidae, a group of marine birds. Greak auks, now extinct, were once common along the shores of western Europe and eastern North America.

GUANINE Reflective substance that gives some fishes their silvery look. Guanine is actually a nitrogenous waste product of the blood. (See also Hypoxanthine.)

SECRETS OF THE DEEP

HALOPHYTE A plant adapted to places of high salinity, such as salt marshes. In other plants, high concentrations of salts in the soil or water have a dehydrating effect, but in halophytes the high salinity of the environment is balanced equally by high intracellular salinities. Uptake of physiologic water, therefore, does not present a problem.

HOLDFAST The base of a marine alga adapted for securing the plant to a solid surface. Unlike the roots of higher plants, holdfasts are not equipped for adsorbing water and nutrients.

HYPOXANTHINE A reflective substance very similar to guanine that gives some fishes a silvery appearance. (See also Guanine.)

INTERTIDAL ZONE Shoreline region between extreme low and high tides.

ISOPOD Crustacean of the Order Isopoda in which the body is compressed from top to bottom (dorsoventrally). Most are marine forms; some can survive out of water for extended periods.

KELP Any of several large seaweeds. Kelps are found only in cool water, usually growing in extensive beds just offshore.

KRILL Whaler's term for crimson, shrimplike crustaceans of the genus *Euphausia*. These two-and-a-half-inch animals are the principal food of many mammals and birds in the short antarctic food chains. Many species of whales, seabirds, and one seal (the crabeater seal) depend on krill directly to sustain them.

LATERAL LINE A line of sensory pores and nerve endings extending along the sides of fishes and a few amphibians. The lateral line detects water movement.

LITTORAL DRIFT Movement of sand along a coast caused by wave-generated currents. Most waves strike a shoreline at an angle, causing sand to move along a coast in the same direction the waves are breaking.

LUCIFERASE Enzyme catalyzing the oxidation of luciferin. (See also Biological luminescence and Luciferin.)

LUCIFERIN A pigment in luminescent animals and plants that produces light when oxidized. (See also Biological luminescence and Luciferase.)

GLOSSARY

LUNAR DAY The interval between two passages of the moon. The lunar day would be the same length as a solar day were the moon not moving; thus, a lunar day encompasses a period of twenty-four hours and fifty-one minutes. (See also Solar day.)

MELON The bulbous forehead region in a cetacean, composed of fibrous and fatty tissue, which is thought to function as an acoustic lens for focusing sound when the animal is using its sonar.

MUTUALISM A symbiotic arrangement between two very different organisms in which both parties benefit. (See also Commensalism, Parasitism, and Symbiosis.)

NAVIGATION The ability of an animal to determine its direction of movement and its location at a given moment with respect to an ultimate destination. All animals that make seasonal or daily migratory journeys can navigate.

NEMATOCYSTS The specialized stinging cells of corals, sea anemones, sea jellies, and their relatives.

NONESCAPEMENT HYPOTHESIS Hypothesis supporting the concept that biological clocks in living organisms are controlled by an outside influence, probably geophysical. This unknown force signals the clock periodically and "sets" it. (See also Biological clock and Escapement hypothesis.)

PACK ICE Ring of low icebergs encircling the antarctic continent. Pack ice is formed from sea ice that breaks free of the continent in spring. The resulting icebergs drift north until they encounter prevailing westerly winds at about south latitude 65 degrees, where their progress is stopped.

PARASITISM A partnership between two organisms in which one lives at the expense of the other. (See also Commensalism, Mutualism, and Symbiosis.)

PENGUIN Flightless marine birds of the Family Spheniscidae. There are six genera and seventeen species of penguins, all of which live south of the equator.

PHEROMONE A class of hormonelike substances that some organisms excrete into the environment. Pheromones are highly selective in the other organisms they affect and the response they elicit.

SECRETS OF THE DEEP

PHYTOPLANKTON Plant plankton. (See also Plankton and Zooplankton.)

PINNIPED Any seal, sea lion, or walrus. From Latin, meaning "winged foot," a description alluding to the long, graceful shape of the flippers.

PLACODERMS Oldest of all vertebrates with jaws. Now extinct.

PLANKTON Tiny animals and plants that drift in the sea. Planktonic organisms form the base of oceanic food chains. (See also Phytoplankton and Zooplankton.)

PLATELET Stack of guanine and hypoxanthine crystals in the skin of a fish. The function of a platelet is to reflect light. (See also Guanine and Hypoxanthine.)

PRIMARY PRODUCERS Microscopic plants that form the first links in marine food chains. (See also Phytoplankton, Plankton, and Secondary producers.)

RHIZOME Underground stem of some perennial, herbaceous plants. A rhizome sends out shoots from its upper end, aiding the plant to colonize new areas. Many dune grasses spread by means of rhizomes.

SALTATION The process by which sand is carried along in the wind.

SALT MARSH Low-lying swampy area found behind sandbars and barrier beaches. Salt marshes were once estuaries. They form where turbulence is reduced, allowing silt and debris to accumulate and support marsh grass. A salt marsh is regularly flooded by tides. (See also Estuary.)

SCHOOL An organized group of aquatic animals swimming together; a shoal. Schooling animals all move in the same direction, keep the same distance from each other, and conduct the same activities in synchrony.

SEA ICE Ice formed from freezing seawater that in winter in the polar zones becomes annexed to the land. In spring the sea ice breaks apart to form drifting icebergs.

SECONDARY PRODUCERS Tiny animals that form the second links in marine food chains. They derive nourishment from the primary producers. (See also Plankton, Primary producers, and Zooplankton.)

GLOSSARY

SESSILE Permanently attached to one spot.

SOLAR DAY The twenty-four hours that it takes the earth to make one complete revolution on its axis. (See also Lunar day.)

SONAR A military acronym for *so*und *n*avigation *r*anging. Biological sonar is used by animals (in the sea, notably by the dolphins and porpoises) to avoid objects and enemies and to find food.

SPECTROGRAM The graphic representation on paper of sound pattern distributions; a "sound picture."

SUN-ARC HYPOTHESIS A theory advanced to explain how birds can navigate using celestial cues. The theory presumes that flying birds can measure the movement of the sun across the sky and extrapolate from it enough of an arc to compare to the "home arc." Experimental evidence shows that the avian eye is highly developed and probably capable of detecting such movements.

SYMBIOSIS An arrangement in which two very different organisms live together. From Greek, meaning "living together." (See also Commensalism, Mutualism, and Parasitism.)

TUBE FEET Specialized feet operated by hydraulic pressure and characteristic of the sea stars, sea urchins, and their relatives.

TURTLE GRASS A sea grass indigenous to shallow tropical waters. Unlike most marine plants, which are algae, turtle grass is a true flowering plant. Named for the large green sea turtles that feed on it extensively.

ZOOPLANKTON Animal plankton. (See also Phytoplankton and Plankton.)

ZOOXANTHELLAE One-celled symbiotic algae growing within the tissues of living corals and some other marine invertebrates.

Selected Bibliography

Adler, Helmut E. (ed.). "Orientation: Sensory Basis." *Annals of the New York Academy of Sciences* 188 (1971):1–408.

Ardrey, Robert. *The Territorial Imperative*. New York: Atheneum, 1966.

Bascom, Willard. *Waves and Beaches*. New York: Anchor, 1964.

Beebe, William. *Half Mile Down*. New York: Harcourt Brace, 1934.

Berrie, A. D., and Visser, S. A. "Investigations of a Growth-inhibiting Substance Affecting a Natural Population of Freshwater Snails." *Physiological Zoology* 36 (1963):167–173.

Buchsbaum, Ralph, and Milne, Lorus J. *The Lower Animals: Living Invertebrates of the World*. Garden City, N.Y.: Doubleday, 1960.

Caldwell, Melba C., Caldwell, David K., and Miller, J. Frank. "Statistical Evidence for Individual Signature Whistles in the Spotted Dolphin, *Stenella plagiodon*." *Cetology* 16 (1973): 1–21.

Cohen, Daniel M. "How Many Recent Fishes Are There?" *Proceedings of the California Academy of Sciences* 38 (1970): 341–345.

Denton, Eric. "Reflectors in Fishes." *Scientific American* 224 (January 1971):64–72.

DeVries, Arthur L. "Freezing Resistance in Fishes." In *Fish Physiology*, Vol. 6, ed. W. S. Hoar and D. J. Randall. New York: Academic Press, 1971.

Eckert, Allan W. *The Great Auk, A Novel*. Boston: Little, Brown, 1963.

Eltringham, S. K. *Life in Mud and Sand*. New York: Crane, Russak, 1971.

SECRETS OF THE DEEP

Grundfest, Harry. "Electric Fishes." *Scientific American* 203 (April 1960):115–124.

Halle, Louis J. *The Sea and the Ice: A Naturalist in Antarctica.* Boston: Houghton Mifflin, 1973.

Halstead, Bruce W. *Poisonous and Venomous Marine Animals of the World,* Vol. 3. Washington, D.C.: U.S. Government Printing Office, 1970.

Hardy, Alister C. *The Open Sea: The World of Plankton.* Boston: Houghton Mifflin, 1956.

Idyll, Clarence P. *Abyss.* New York: Thomas Y. Crowell, 1971.

Johansen, Kjell. "Air-breathing Fishes." *Scientific American* 219 (April 1968):102–111.

Kalle, K. "The Problem of the Gelbstoff in the Sea," *Oceanography and Marine Biology Annual Review* 4 (1966):91–104.

Kellog, Winthrop N. *Porpoises and Sonar.* Chicago: University of Chicago Press, 1961.

Kuenen, P. H. "Sand." *Scientific American* 202 (April 1960): 94–110.

Mariscal, Richard N. "Behavior of Symbiotic Fishes and Sea Anemones." In *Behavior of Marine Animals,* vol. 2, ed. Howard E. Winn and Bori L. Olla. New York: Plenum, 1971.

Marshall, N. B. *The Life of Fishes.* Cleveland: World, 1966.

Matthews, G. V. T. *Bird Migration.* 2d ed. Cambridge Monographs in Experimental Biology No. 3. Cambridge, England: Cambridge University Press, 1968.

McElroy, William D., and Seliger, Howard H. "Biological Luminescence." *Scientific American* 207 (June 1962):76–89.

Moore, Hilary B. "Aspects of Stress in the Tropical Marine Environment." *Advances in Marine Biology* 10 (1972):217–269.

Murphy, Robert Cushman. "The Oceanic Life of the Antarctic." *Scientific American* 207 (March 1962):186–210.

National Marine Fisheries Service. *Fishery Statistics of the United States 1968.* Statistical Digest 62. Washington, D.C.: U.S. Government Printing Office, 1971.

National Marine Fisheries Service. *Fisheries of the United States, 1974.* Current Fishery Statistics No. 6700. Washington D.C.: U.S. Government Printing Office, 1975.

SELECTED BIBLIOGRAPHY

Norman, J. R., and Greenwood, P. H. *A History of Fishes.* Rev. ed. New York: Hill and Wang, 1963.

Palmer, John D. *Biological Clocks in Marine Organisms.* New York: Wiley, 1974.

Ranwell, D. S. *Ecology of Salt Marshes and Sand Dunes.* London: Chapman and Hall, 1972.

Schlichter, D. "Chemische Tarnung. Die stoffliche Grundlage der Anpassung von Anemonenfischen an Riffanemonen." *Marine Biology* 12 (1972):137–150.

Schroeder, Robert E. *Something Rich and Strange.* New York: Harper & Row, 1965.

Shaw, Evelyn. "The Schooling of Fishes." *Scientific American* 206 (June 1962):128–138.

Smith, F. G. Walton. *Atlantic Reef Corals.* Miami, Fla.: University of Miami Press, 1971.

Teal, John, and Teal, Mildred. *Life and Death of the Salt Marsh.* New York: Ballantine, 1969.

Tickell, W. L. N. "The Great Albatrosses." *Scientific American* 223 (May 1970):84–93.

Todd, John H. "The Chemical Languages of Fishes." *Scientific American* 224 (May 1971):98–108.

Wilson, Edward O. "Pheromones." *Scientific American* 208 (May 1963):100–114.

Wood, Forrest G. *Marine Mammals and Man: The Navy's Porpoises and Sea Lions.* Washington, D.C.: Luce, 1973.

Index

Page numbers in parentheses indicate illustration legends.

INDEX

INDEX

INDEX

INDEX

About the Author

Stephen Spotte is Vice President and Director of Aquariums for Aquarium Systems, Inc., in charge of Aquarium of Niagara Falls (New York) and Mystic Marinelife Aquarium in Mystic, Connecticut. He has been Curator of the New York Aquarium and Osborn Laboratories of Marine Science and before that was, successively, Curator of Exhibits, General Curator, and Director of Aquarium of Niagara Falls. Mr. Spotte, who holds a Bachelor of Science degree from Marshall University, has also worked as a field biologist for the U.S. Army Corps of Engineers, as a professional diver in the Caribbean, and as a free-lance writer and photographer. He is the author of two books on fish culture and numerous popular and scientific articles. He makes his home in North Stonington, Connecticut.

About the Illustrator

Gordy Allen was born and raised in Baltimore, Maryland. Since his graduation from high school in 1972 he has made a living as an oysterman and duck hunting guide on the Chesapeake Bay and has trapped whistling swan and other waterfowl for the Fish and Wildlife Service in Maryland. Mr. Allen attended the Maryland Institute for one year and began illustrating books one year ago. *Secrets of the Deep* is his third book. He is presently living on the Delaware River in Pennsylvania.